Happy Valentine's Day
Sweetheart!
Thought this would bring
back some great and funny memories
for you!
Luv,
Linda
xx oo

IMAGES
of America

BENSON'S
WILD ANIMAL FARM

A pair of male lions enjoyed the warm sunshine in their outside exhibit at Benson's Wild Animal Farm. (Author's collection.)

IMAGES
of America

BENSON'S
WILD ANIMAL FARM

Bob Goldsack

ARCADIA
PUBLISHING

Copyright © 2011 by Bob Goldsack
ISBN 978-1-5316-4873-2

Published by Arcadia Publishing
Charleston, South Carolina

Library of Congress Control Number: 2010932743

For all general information, please contact Arcadia Publishing:
Telephone 843-853-2070
Fax 843-853-0044
E-mail sales@arcadiapublishing.com
For customer service and orders:
Toll-Free 1-888-313-2665

Visit us on the Internet at www.arcadiapublishing.com

CONTENTS

ACKNOWLEDGMENTS

Sincere appreciation is extended to the following people who provided information and photographs that made this historic journey possible: Arthur Provencher, Benson's owner 1979–1987; Ruth Parker, Hudson Historical Society; Esther McGraw, Benson volunteer chairperson; Cindy (Martin) Provencher and Bret Bronson, former Benson employees and performers; Kenneth Matthews, Friends of Benson's president; Dareen Gannice, freelance writer; and Steven Klein, who directed us to various Benson Web sites.

A special acknowledgment is also given to the late Benson's wild animal trainer Joseph Arcaris, who provided many Benson photographs, identifications, and stories.

All images without a courtesy line are from the author's files and photographs.

INTRODUCTION

Benson's Wild Animal Farm (later changed to Park in 1988) was a New England landmark for most of its years in business. The attraction began shortly after John T. Benson purchased property in 1922 for a place to quarantine imported animals before they were sold to zoos and circuses. Most circuses and animal trainers at that time obtained their wild animals from Benson, who served as a dealership for the Hagenbeck Company of Hamburg, Germany, the largest dealer in wild animals during that time. Benson learned his trade while traveling with the Bostock and Wombwell Circus in England, and he accompanied it to the United States in 1890.

In an effort to discourage his Hudson neighbors from requesting permission to enter the farm and watch the animals, Benson decided to charge a minimal admission. His idea backfired, however, and soon people from all over New England were stopping by to see the wild beasts. An astute businessman, Benson quickly realized the potential of his animal farm. He began landscaping and building areas for the animals to roam free, plus barns to keep them warm during the cold winters.

By 1934, the parking lot could accommodate 5,200 cars, and most weekends there was an overflow. Area people have remarked that on any given Sunday they could walk through the parking lot and see license plates from just about every state in the union. For a number of years starting in the 1930s, the Boston and Maine Railroad ran a special round-trip *Jungle Train* from Boston's North Station to the Hudson depot every Sunday during the entire season. During Benson's ownership, many famous circus animal trainers appeared at the facility, and from its start until its closing 61 years later, animal acts of every type were presented on a daily basis.

In a neighborly gesture, Benson, in the mid-1930s, announced that Hudson residents were exempt from having to pay admission. Due to World War II restrictions, Benson decided to close his operation for the duration of the war, starting with the summer of 1942. Unfortunately, Benson died in 1943. A consortium of four Boston men purchased the land, contents, and title in 1944 and reopened the farm in 1945. The group divested itself of the animal trading and selling part of the business and put all their efforts into the attraction itself.

The word "farm" in its title and corporation papers allowed children 14 years and older to work at the attraction in accordance with child labor laws. As fast food operations were well in the future, Benson's was the only place in town where, according to child labor laws, teenagers could earn some money. Generations of Hudson teens worked at Benson's during their high school years.

As the members of the consortium gradually passed away, less and less money was reinvested and the attraction started a slow slide into visible neglect. Arthur Provencher, a successful area businessman and dedicated animal lover, purchased the animal farm from the estate of the last partner in 1979. Provencher poured money into the infrastructure and attractions, acquiring added animal species and eventually a number of mechanical rides. He also increased advertising and promotion, and it was estimated that attendance increased almost 70 percent his first year of ownership.

Unfortunately, it was a time when long-term financing was impossible and prevailing short-term loans ran at 24 percent interest. Within a couple of years he was indebted to the bank for $2.4 million and another $1.1 million to other creditors. It became necessary to sell his successful business and property to keep Benson's operating, but it was not enough to keep the wild animal farm solvent. He then sold off some of the unused acreage to satisfy the $2.4 million owed to the bank.

Advisors suggested raising admission prices but he tried to continue without a price increase. Although attendance continued to climb each year, the revenue was not enough to sustain expenses and the park closed at the end of the 1987 season.

One

HUDSON, NEW HAMPSHIRE

Hudson, New Hampshire, is a picturesque New England community of the type so well illustrated by Norman Rockwell. Only since the 1960s has the town experienced a rapid growth in industry, retail establishments, and home construction. Beautiful renovated old farmhouses are still a part of the community, which sprawls over a rather large area.

Alvirne, the local high school, still offers agriculture along with its vocational courses. A miniature farm is maintained on the spacious school grounds, where students can earn a hands-on farm experience. In addition to a small dairy herd and other farm animals, a tree farm is maintained and various crops are produced.

The city of Nashua, located across the Merrimack River from Hudson, has always been on the major road heading north through the central part of the state. It was originally Route 3 and then renamed the Everett Turnpike. It then becomes Route 293 for a short distance before joining Route 93, just past Manchester, New Hampshire. Nashua was also a major railroad center for the area. For this reason, Nashua was often misrepresented as the home of Benson's on postcards and other literature.

But that never bothered Hudson, as it basked in its fame as it hosted visitors from all over the United States and on occasion the world. Benson's was the attraction that provided the community its own unique identity. Weekend traffic became hectic as automobiles, buses, and railroad coaches deposited thousands of wide-eyed tourists who enjoyed a day in the country viewing exotic animals in their corrals and watching trained animals perform circus-style acts at various locations throughout the "Strangest Farm on Earth."

For years, Benson's provided summer employment for local teenagers, and during the John T. Benson years, free admission became a privilege enjoyed by Hudson residents. Esther McGraw, a lifelong Hudson resident and former Benson employee, recalls out-of-towners going through the telephone book to pick out a name and address of a Hudson resident and use it to gain free admission.

Longtime Hudson residents still enjoy reminiscing about the wonderful times they spent at the attraction. Thousands of photographs, slides, movies, and souvenirs of Benson's still remain in Hudson homes, and every once in awhile they are brought out and happy times are remembered.

BENSON'S ANIMAL FARM, HUDSON, N. H.

The historical John T. Benson home was torn down by the State of New Hampshire after the park closed. The state took the property by right of eminent domain for the purpose of wetland mitigation. (Courtesy Hudson Historical Society.)

Two

John T. Benson

John T. Benson was born in 1879 in Yorkshire, England. Two different stories exist as to how he became interested in animals. One has him growing up in his father's animal show touring the British Isles, while the storybook version has him running away from home at the age of eight to join Bostock and Wombwell's traveling circus. It has been established that Benson did troupe with the Bostock and Wombwell Circus and accompanied the show when it came to the United States in 1890. His job was to perform with a wrestling lion.

Benson soon became the U.S. representative of the Hagenbeck Company of Germany, the largest dealer in wild animals at that time. He sold wild animals from all over the world to zoos and circuses throughout the country for over 40 years. His Hagenbeck association ended with the outbreak of World War II.

He also spent time working in the zoo departments of Lexington Park in Lexington and later at Norumbega Park in Waltham. Both parks were located in Massachusetts. A very important stepping stone in his career occurred in 1911 when he helped establish and organize the Franklin Park Zoo in Boston.

Wild animal noises and smells emanating from the Hoboken, New Jersey, wharf warehouse where Benson kept his quarantined animals brought protests from neighbors, and he was ordered to relocate. The 150-acre former farm property was purchased in 1922 and began operating under the Interstate Animal Farm title.

Benson acquired quite a reputation as an accomplished animal trainer and a top authority in the wild animal business. He also became very well known in political circles and Pres. Theodore Roosevelt invited him as a companion on a trip to Africa to hunt exotic animals. He also traveled with Roosevelt on a special hunt in the Black Hills of South Dakota. As a token of his appreciation, Roosevelt presented him with two beautiful Great Dane puppies, Freda and David, whom Benson raised and enjoyed for many years.

John T. Benson passed away in 1943, leaving several heirs who were never active in the park. A nephew, Ron MacKenzie, did work several years for his uncle.

John T. Benson enjoyed a rickshaw ride during a safari trip to India, Africa, and Indochina in 1915. The trip was commissioned by the Carl Hagenbeck Company of Hamburg, Germany. Benson captured hundreds of animals and birds for the company.

Here is an early John T. Benson advertisement placed from his newly acquired New Hampshire farm. The advertisement appeared in *Billboard* magazine, the weekly show business trade paper that covered all forms of show business.

**If It's . . . .
WILD ANIMALS
REPTILES, BIRDS OR MONKEYS
. . . . You Want
WRITE OR WIRE JOHN T. BENSON, NASHUA, N. H.**

An elderly John T. Benson was photographed with one of his little friends as they posed under an apple tree at Benson's Wild Animal Farm in the late 1930s. (Courtesy Arthur Provencher.)

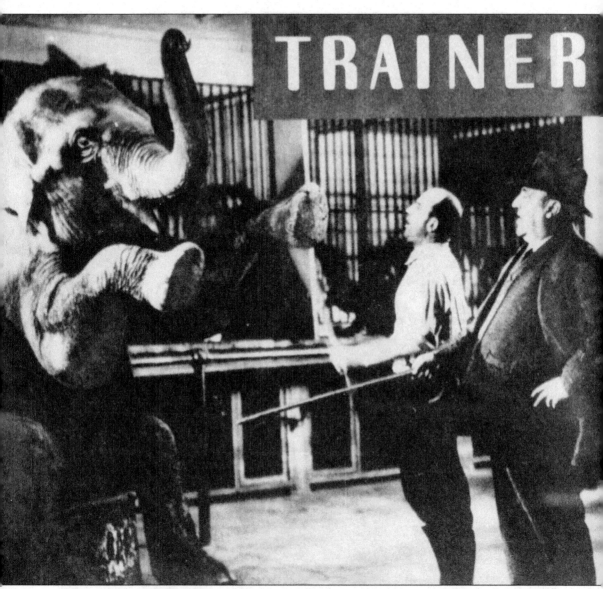

In a magazine article, it was said that "John T. Benson is the man who trains the lions, tigers, elephants and other animals seen in circuses and vaudeville houses all over the United States."

Three

BENSON MOVES

TO NEW HAMPSHIRE

Benson had an urgent need to vacate Hoboken due to the complaints of locals regarding the noises and smells of his animals. The convenience of receiving animal shipments in Boston suddenly had Benson seeking property in a remote New England area. Made aware by show business friends that the Hudson property was for sale, he purchased what had been known as the Interstate Fruit Farm, and later a health farm, for a reasonable price.

The 150-acre property was acquired in 1922 and operated under the Interstate Animal Farm title. A small portion of the land was quickly converted into corrals, cages and pens. The spacious acreage enabled Benson to establish quarantine zones for animals arriving in the United States, an area to showcase animals to be sold to zoos, and a facility to train animals that would be sold to circuses and vaudeville performers.

Nearby Hudson residents soon became curious when they noticed trucks carrying strange creatures arriving at the farm from which many strange noises could be heard, especially at night. It did not take long for the inquisitive to ask permission to enter the premises for closer views of the animals. Figuring he would discourage these interruptions, Benson started to charge a small fee to enter his farm. This ploy did not work, as more and more visitors began appearing at his gate. Always an entrepreneur, he immediately set aside 4 acres of pens and cages for visitors to view his stock. As attendance increased, ice cream, popcorn, soda stands, and a lunch area were available. Pony and elephant rides, carts pulled by goats, and other animal rides became available for small fees.

During this time, animals were being trained and sold or leased to movie studios. The first chimpanzees that appeared in the *Our Gang* movies were provided by the Hudson farm. He also received a contract to stock the Adirondack Mountains with elk, moose, and bears under the direction of the New York Fish and Game Commission. During the 1930s, Benson was also listed as one of the world's largest distributor of goldfish, second only to Japan.

Each year the visitor count grew larger and larger as a host of new attractions would appear. Between Sunday April 24 and May 1, 1932, attendance reached a high of 3,000 paid admissions. John T. Benson knew he was doing something right.

A boatload of camels is shown as it arrived at John T. Benson's Hoboken, New Jersey, pier in 1918. The camels were unloaded and kept in quarantine at the Hoboken warehouse until sold to zoos and circuses. (Courtesy Arthur Provencher.)

The wild animal noises and odors emanating from these lion cages at Benson's Hoboken warehouse brought protests from residents living near the dock area, and the town administrators ordered Benson to vacate the premises. (Courtesy Arthur Provencher.)

John T. Benson believed that the admission charge would discourage the townspeople from asking to see the animals at his farm. In 1926, he set aside only a 4-acre area filled with small cages and pens for the animals. (Courtesy Arthur Provencher.)

As the development of his attraction grew, Benson added performances by a variety of his animals. In this photograph, a crowd was gathering for the chimpanzee Kindergarten Tea Time performance. (Courtesy Arthur Provencher.)

BENSON'S WILD ANIMAL FARM, NASHUA, N. H

A young boy was photographed fishing in a small pond on the Benson property. At one time Benson raised, packaged, and distributed goldfish to the world. According to his nephew Rod MacKenzie, Benson was listed second only to Japan in this industry.

Monkeys and chimpanzees have been regarded as great performing comedians. People always enjoyed watching them perform at Benson's Monkey Circus. Dressing them in cute outfits and training them to do amazing antics was a natural crowd pleaser. (Courtesy Arthur Provencher.)

On a Sunday morning in 1926, visitors stream through the Benson Wild Animal Farm entrance gate. The exit sign on the left read, "What would P. T. Barnum say if he could have seen what you saw here today?" (Courtesy Arthur Provencher.)

Adorable ponies delighted audiences several times a day as they performed unusual tricks. (Courtesy Arthur Provencher.)

This photograph was taken in the late 1920s and illustrates how well people dressed when they left their homes during that time period. The crowd had gathered to watch a monkey or chimpanzee show, which was one of many animal presentations provided throughout the day. Benson's was

noted for the many flower beds and plantings, mixed in with shade trees throughout the grounds. The striped umbrella tables were placed where visitors could obtain a cool drink or a bite of lunch while sheltered from the sun. (Courtesy Arthur Provencher.)

Benson's Ark was a popular site of a reenactment of Noah's story from the Bible. A parade showing two of each species was held daily. This area was what today would be known as the petting zoo. (Courtesy Arthur Provencher.)

A year or two later, Benson's Ark featured a fenced area where visitors could watch farm animals at leisure. Stairways were installed at each end of the Ark to allow passage through the barn to view additional farm animals.

Four

BENSON EXPANDS
THE FARM

John T. Benson knew his new venture was successful, and starting in 1932 he began adding many new features and attractions each season.

It was not unusual to see a pig wearing a dress and pulling a cart with a sign reading "Bringing Home The Bacon," or a rooster dressed up in pants. One time, during an election year, a donkey wandered throughout the farm wearing a sandwich board sign that read "Vote for Alf Landon." A skeleton of former Ringling Brothers' famous elephant Old John and a famous horse were on display along with two Egyptian mummies that Benson was said to have brought back from one of his safaris.

Advertising announced that in 1933, Benson's Permanent Wild Animal Circus would be presented on a daily basis. Reports claimed that nearly 100 animals were to be presented in the various acts throughout the property. Frank Wooska handled wild Bengal tigers and Bruno Radtke worked with African and mountain lions. Charles Barry, formerly with the Al G. Barnes Circus, presented trained pigs, goats, and antelope. Robert MacPherson, formerly of the Hagenbeck-Wallace Circus, pleased the audience with his Animal Kindergarten. Capt. Fred Leonard cued the liberty horses and ponies while Carl Neuffer and Herbert Riddle put the elephants through their paces. Hilda Miller lectured as she showed off a variety of snakes while the chimpanzees of Margaret Thompson and Mrs. Bruno Radtke really pleased their audiences. Over the following years, many well-known circus performers appeared in the Benson circus.

John T. Benson has been described by those who knew him as a hearty, likeable chap who was always ready with a joke or funny story. During Benson's reign, he hired top landscapers who kept the visitor areas alive with beautiful and colorful plantings during each season. It has also been said that part of Benson's success was his ability to surround himself with top assistant managers and department heads. He was a stickler for appearance and provided his help with uniforms.

At the end of each summer, kids who worked all season were presented with a $18.75 gift certificate, redeemable at a Nashua clothing store. John T. Benson felt children should start the school year with a new set of clothes.

Named after the famous aviator, Lindy the chimpanzee pulled his miniature airplane across the treetops on a wire. His motivation was peanuts deposited in a bucket hanging on each end of the wire. (Courtesy Arthur Provencher.)

Several trucks were used to transport animals to and from the Nashua freight station and to pick up arriving animals at Boston Harbor. The colorful trucks, well decorated with signs and artwork advertising the attraction, were also used for area parades. (Courtesy Hudson Historical Society.)

John T. Benson poured a mug of beer for visiting dignitaries in his beer garden after Prohibition ended in the United States. (Courtesy Arthur Provencher.)

Benson quickly realized the financial potential of his attraction and he began soliciting outings by civic, corporate, and other organizations that involved catered barbecues. In 1931, he hosted the annual convention of the Circus Fans of America. (Courtesy Joseph Arcaris.)

The bear cubs were presented daily in a large, sunken, cement oval area known as the bear pits. The little bruins would stand on their hind legs and beg for food. (Courtesy Arthur Provencher.)

Each year, Native Americans in Manitoba would capture a half-dozen bear cubs and ship them to John Benson. They were displayed in a large cylindrical space where they frolicked, much to the enjoyment of onlookers.

Many Benson animal acts were rented to circuses and other shows in the off season. Trainer Franz Wooska put four Benson tigers through their paces while appearing in the musical *Jumbo* at the New York Hippodrome from November 1935 through March 1936.

Chief Young Thunder Cloud and his Indian Trading Post gift shop were very popular during Benson's ownership. (Courtesy Esther McGraw.)

John T. Benson guided his elephant as he provided a special ride to three distinguished-looking gentlemen. (Courtesy Arthur Provencher.)

Betsy, originally trained by John Benson, was kept very busy around the farm. In this photograph she carries 10 children and is supervised by elephant man Carl Neuffer. She also performed and helped move fallen trees and other heavy objects around the farm.

World-famous tiger trainer and presenter Mable Stark performed daily during the 1940 season at Benson's. Her thrilling act was a longtime attraction with Al G. Barnes and other major circuses of the 1920s and 1930s.

This 1937 handbill advertised the "Jungle Train," which made a round-trip excursion from Boston's North Station to Benson's every Sunday during the season.

This *Jungle Train* arch appeared in Boston's North Station during the summer season each year. A combination ticket covered both the train ride and admission as the farm billed the railroad for the price of each passenger. (Courtesy Arthur Provencher.)

This was the Hudson railroad depot, where the *Jungle Train* deposited its excursion riders on Sunday mornings. The depot was located across the road from Benson's and behind the Grange Hall, which still exists today. Exactly at 4:00 p.m., the train whistle would pierce the air, alerting its riders that departure time would be in half an hour.

Just Arrived from India

Two baby elephants from India were obtained through John T. Benson's association with the Carl Hagenbeck Company of Hamburg, Germany, the largest dealer in wild animals at that time.

The Wonder Place of New England!

FREE PARKING For Thousands of Cars

A DELIGHTFUL RIDE FROM ANYWHERE

THE NATURE LOVERS' PARADISE!

This publicity photograph of two nurses attending to baby chimpanzees sitting in high chairs was used as part of a folder sold at the souvenir building. This was part of the Kindergarten Tea Time presentation.

A trained high school horse bends into a most difficult and unnatural position as it bows to the audience—a common ending for horse performances.

Rajah was a magnificent tiger who spent many years performing at Benson's. Tigers, lions, chimpanzees, monkeys, horses, and elephant acts were the most popular acts over the years. Birds, bears, zebras, and ponies were also enjoyed. (Courtesy Eve Wright.)

This Benson live pony merry-go-round was probably the only one-of-a-kind animal ride found anywhere. The overhead sweep with the rounding boards was so different from the usual ground-mounted sweep used for pony rides. (Courtesy Hudson Historical Society.)

Another unusual Benson ride was the merry-go-round powered by live spotted deer. The variety of live animal rides gave Benson's Wild Animal Farm a great reputation with people from all over the country. (Courtesy Arthur Provencher.)

JOHN T. BENSON
THE MASTER MIND OVER THE WILD ANIMAL KINGDOM
INVITES YOU TO VISIT
THE STRANGEST FARM ON EARTH!
5 MINUTES FROM NASHUA, N.H. UNION STATION

A CHILDREN'S PARADISE
An Education - Instructive - Amusing
Fun and Health Producing Institution
A Special Program for the Children is going
on at intervals during the day

A Few of the Great Diversified Programs

THEY CAN RIDE

BETSY, the most valuable elephant in the world

on PONIES, specially trained for safety

Indian Antelopes

Merry-Go-Round

THE COMICS OF THE FARM ALWAYS BUSY
The Airplane monkeys – Liza, the itinerant donkey with Liz, his Siamese companion, – Mr. Gander, always on parade – Jimmy, "the Porkine Express Runner" – Baby ponies and donkeys.

THE KIDDIES' TOY SHOP — A complete line of miniature imitations of all animals on the Farm.

THE RECREATION GROUNDS FOR ALL KINDS OF SPORTS FOR BOYS AND GIRLS, AND FOR MYSTERY LOVERS—THE MAZE!

THE PICNIC GROUNDS, delightfully appointed, with tables, benches in shady spots for the comfort and convenience of family and group parties.

WATER PIPED from the constant FLOWING SPRINGS IS ON TAP ALL OVER THE FARM.

FOR CONVENTIONS — Outings, picnic parties, family gatherings — it is known as "THE WONDER PLACE OF NEW ENGLAND."

IN THE GARDEN—A Cafe where lunches and drinks are served at reasonable prices-Refreshment stands are located at various parts of the Farm.

Located five minutes from Nashua Union Station, seven minutes from Nashua Main Street and fifteen minutes from the Nashua Airport.

Good roads from all parts of New England.

Open from 10 a.m. to 6 p.m. daily until Thanksgiving.

47 ACRES OF FREE PARKING SPACE **47**

TELL YOUR FRIENDS HOW TO GET TO

BENSON'S
WILD ANIMAL FARM

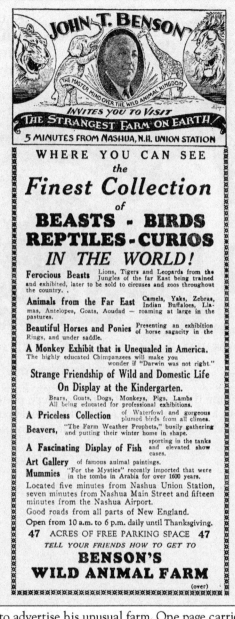

JOHN T. BENSON
THE MASTER MIND OVER THE WILD ANIMAL KINGDOM
INVITES YOU TO VISIT
THE STRANGEST FARM ON EARTH!
5 MINUTES FROM NASHUA, N.H. UNION STATION

WHERE YOU CAN SEE
the
Finest Collection
of
BEASTS - BIRDS
REPTILES - CURIOS
IN THE WORLD!

Ferocious Beasts Lions, Tigers and Leopards from the Jungles of the far East being trained and exhibited, later to be sold to circuses and zoos throughout the country.

Animals from the Far East Camels, Yaks, Zebras, Indian Buffaloes, Llamas, Antelopes, Goats, Aoudad — roaming at large in the pastures.

Beautiful Horses and Ponies Presenting an exhibition of horse sagacity in the Rings, and under saddle.

A Monkey Exhibit that is Unequaled in America.
The highly educated Chimpanzees will make you wonder if "Darwin was not right."

Strange Friendship of Wild and Domestic Life On Display at the Kindergarten.
Bears, Goats, Dogs, Monkeys, Pigs, Lambs All being educated for professional exhibitions.

A Priceless Collection of Waterfowl and gorgeous plumed birds from all climes.

Beavers, "The Farm Weather Prophets," busily gathering and putting their winter home in shape.

A Fascinating Display of Fish sporting in the tanks and elevated show cases.

Art Gallery of famous animal paintings.

Mummies "For the Mystics" recently imported that were in the tombs in Arabia for over 1600 years.

Located five minutes from Nashua Union Station, seven minutes from Nashua Main Street and fifteen minutes from the Nashua Airport.

Good roads from all parts of New England.

Open from 10 a.m. to 6 p.m. daily until Thanksgiving.

47 ACRES OF FREE PARKING SPACE **47**

TELL YOUR FRIENDS HOW TO GET TO

BENSON'S
WILD ANIMAL FARM
(over)

These were two early handbills used by Benson to advertise his unusual farm. One page carries the address as Nashua, but the location of Hudson Center appears in small type. (Courtesy Hudson Historical Society.)

MOTHER, YOU CAN'T MAKE ME BELIEVE THE STORK BROUGHT THAT BABY.

A SCENE AT BENSON'S WILD ANIMAL FARM, NASHUA, N. H.

During the first half of the past century, postcards were extremely popular. Any time people traveled or visited an attraction, they would purchase postcards and mail them to relatives and friends. Many Benson postcards remain in various collections today.

In the late 1930s, following the repeal of Prohibition, John T. Benson built this attractive and picturesque beer garden and restaurant. It presented a nice atmosphere for those seeking some refreshment or food. (Courtesy Arthur Provencher.)

This is another view of Benson's German beer garden and restaurant in the late 1930s. Unlike most attractions, visitors were allowed to bring their own picnic lunch. (Courtesy Arthur Provencher.)

The title of head cook was a prestigious designation at Benson's. Carrie Stevens was Benson's official head cook in the 1940s. (Courtesy Lucielle Boucher.)

Five

NEW OWNERSHIP

Following John Benson's death, a consortium of four Boston men, Raymond Lapham, Walter Brown, Harry Collins, and Charles Keene, purchased the attraction. Under the corporate name of Boston Garden Corporation, the group owned the Boston Garden, Boston Arena, and a chain of Boston gas stations. Brown would later become the owner of the Boston Celtics professional basketball team.

The new management quickly divested itself of the animal trading and selling part of the business as it planned to put all its efforts into the attractions itself. World War II was still raging when the purchase took place. Once peace arrived, the farm reopened for the 1946 season. New picnic pavilions were installed to accommodate increased business in annual picnics and outings. School spring outings were enjoyed by schoolchildren from the Boston area and towns throughout Massachusetts, Maine, and Rhode Island in addition to New Hampshire. By 1954, the annual attendance reached a half-million visitors, with general admission at 60¢ for adults and 30¢ for children.

Joe Arcaris worked with lions and tigers while Carl Neuffer presented three elephants. There were Fred Pitkin's ponies and George Marshall's chimpanzees. A fire engine, with its cab built on a Crosley chassis, gave a ride the youngsters just loved.

In 1964, Silvers Madison from the Clyde Beatty Cole Brothers Circus took over the elephant department. That same year, Roland Tiebor and his sea lions and seals began their long association with the farm. Harriett Beatty, daughter of Clyde Beatty, spent a season presenting her eight lions in the big cage. Betsy, John Benson's pet elephant who spent almost all her life at the farm, died of natural causes in 1971 and was buried on the grounds.

During the late 1960s and through the 1970s, little money was reinvested, as only Raymond Lapham remained of the original four owners. The facility started to show signs of neglect, and the property steadily passed into poor physical condition. Only one animal act was being staged each day. In turn, attendance began a steady decline.

Raymond Lapham died in 1976, and Benson's continued its downward spiral.

An important part of the Benson experience was visiting the attractive aviary stocked with hundreds of birds and waterfowl species. Large stones and boulders found on the property were used to build the background at this exhibit.

Baby leopards enjoyed sunning themselves on the lion statue. Concrete statues of a variety of animals could be found throughout the grounds. These included lions, camel, and the wishing fox messenger of Inari, god of rice and idol of expectant mothers in Japan. (Courtesy Eve Wright.)

The main entrance sign announced the opening date for the 1947 season. This was the second season for its owner consortium, which purchased the attraction in 1944 and reopened it after World War II in 1946.

In 1948, Joe Walsh presented his lions and tigers in the big steel arena. Hudson native Ray Parker worked during his high school summers in 1949–1951 and took many excellent slides of the exhibits, animals, and the workers.

Elephant Betsy carried the flag in a patriotic pose in 1954. Gordon and Harold Moore worked with the horses and ponies during their high school years. Gordon had the opportunity to help train five spotted ponies.

Three attendants of the reptile houses handled a variety of large boa constrictors and pythons in their charge. There were three reptile houses, one with a large glass front that housed nine boa constrictors and two pythons.

Matched horses presented an unusual equine long mount. This is an act usually performed by elephants in the circus. The horses were practicing their act during the winter, as snow is on the ground.

Although this stage is shown at a time when it was a playhouse for the chimpanzee act, it was also used for various other acts. In 1954, George Marshall presented the chimpanzees act and was assisted by Ray Haverfield.

Three baby elephants imported from India performed for visitors several times a day. In 1954, Carl Neuffer worked the elephants. Silvers Madison, a veteran circus trainer, later took over the elephant herd. (Courtesy Eve Wright.)

Fred Pitkin's eight matched palominos comprised a great act in the 1950s. Pitkin also presented an act using one pony that could do addition and subtraction, along with two other ponies that rocked back and forth on a seesaw.

The famous hand-carved and painted totem pole, presented to John T. Benson by a Native American tribe in the northwest coast of the United States, stood for decades at the farm.

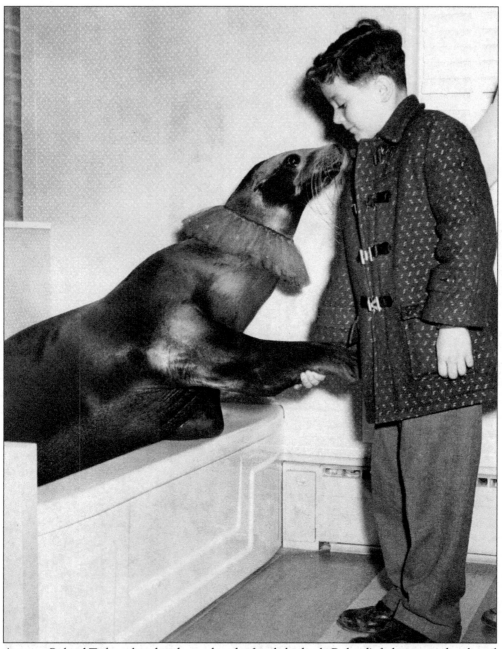

A young Roland Tiebor played with a seal in the family bathtub. Roland's father trained seals and sea lions at the family homestead in North Tonawanda, New York. (Courtesy Esther McGraw.)

As a teenager Tiebor continued to work his sea lion and seal act. Prior to Tiebor's debut, Red Morris appeared with two eight-year-old seals and one sea lion in 1964. (Courtesy Esther McGraw.)

This circus-style poster was used to advertise Roland Tiebor's troupe of educated seals and sea lions. It was only fitting that Benson's, so long associated with the circus, should adopt a circus-style advertising poster. (Courtesy Esther McGraw.)

Roland Tiebor cues his sea animals through a very popular watery performance at Benson's. It is interesting to note that the once-popular circus seal and sea lion acts and their trainers are a rarity today.

Esther McGraw worked for both John T. Benson and Vera Lovejoy, who served as manager under the second ownership. Here Esther drives a tractor delivering supplies in 1969. Worker Julie Breault rides on the back of the tractor. (Courtesy Esther McGraw.)

Visitors view the dromedaries, which are one-hump camels, in their corral. At one time, Benson's camel rides were very popular.

As youngsters in the 1960s, siblings Stasia and Tony Grzesik enjoyed the kiddie boat ride at Benson's. (Courtesy Stasia Dionne.)

Happy parents and children enjoyed the *Jungle Flyer* train as they circled the zoo property. Miniature trains have always been very popular in amusement parks. (Courtesy Stasia Dionne.)

The merry-go-round has always been a favorite ride for children and adults alike. Mother Lillian Grzesik stands nearby as Stasia Grzesik (left) and Susan Medina pretend they are riding real live horses. (Courtesy Stasia Dionne.)

Stasia Grzesik romps along one of the paved paths that led to the many animals and attractions offered throughout Benson's. (Courtesy Stasia Dionne.)

The *Jungle Flyer*, a miniature train installed during John Benson's ownership, toured riders around the grounds. The train had to be replaced when Arthur Provencher purchased the attraction.

This aerial view of the farm was taken in the 1960s. Many of the buildings, pens, and other sites are visible when a magnifying glass is used to study the photograph. (Courtesy Hudson Historical Society.)

A toucan from South America was one of the many bird species from around the world that were seen at Benson's. (Courtesy Eve Wright.)

Deborah and Bruce Boucher were fascinated as they watched a Moby Dick boat float along the waterway. During the Lapham era, several kiddie rides were introduced. (Courtesy Lucielle Boucher.)

Roland Boucher and his children Deborah and Bruce rest as they posed for a photograph to remember their memorable day at Benson's. (Courtesy Lucielle Boucher.)

The fire engine ride was very popular with youngsters as it drove around the upper concourse in 1964. The fire engine was built from an old Crosley automobile. (Courtesy Lucielle Boucher.)

Bruce and Deborah Boucher enjoyed climbing on the concrete camel. The cement animals that were scattered around the grounds served as nice statues, resting spots for tired walkers, and a climbing playground for children. (Courtesy Lucielle Boucher.)

The vintage Allan Herschell merry-go-round carried 30 horses and four benches during its long tenure at Benson's. There was always a line waiting to ride this popular riding device.

Mouflon longhorn sheep were imported from Sardinia for display at the farm. In 1963, nine Himalaya bears were flown in from Tibet. During the 1970s, the number of animals greatly diminished when the Lapham ownership discontinued reinvesting any money in the farm. (Courtesy Eve Wright.)

A birdcage structure on the upper concourse held a sign that read, "Welcome to the Strangest Farm on Earth." The fire engine can also be seen, along with several buildings.

A good-sized crowd enjoyed four beautiful spotted ponies performing part of their liberty act by jumping over barriers. (Courtesy Esther McGraw.)

John T. Benson, Inc.

Wild & Domestic Animals, Birds & Goldfish

Nashua, New Hampshire

(Hudson Centre)

April 16, 1940

Mr. Joseph Arcaris,
c/o Clyde Beatty Zoo
Fort Lauderdale, Florida

Dear Sir;

    Have your letter of April 12th with photograph enclosed and the note also enclosed. I am afraid you have got the wrong impression about the cost of an animal farm. There are 9 animal farms started to my knowledge in the past 7 years and they have all closed up.

    Now, the situation is as follows: Mable Stark is working the tigers and I feel that I would like to have a man to work the 3 lions which are in excellent condition and easy to work. According to what you say, you can probably work them twice a day depending on what we have to do. You know we close about 6 o'clock every night, you have to help to clean the cages in the morning and cut the meat, work the 3 lions and help Mable Stark during her act.

    The regulations here are entirely different than what you have been accustomed to in the past three years. You know we have the best class of people come here and we do not use any show people's methods at all, so if you think you can handle the job according to the terms you state in your letter, give the management where you are a week's notice and be sure to leave him right so you will not be afraid to meet them again and be able to work for them again. If you can handle the above and cooperate with Mable Stark, I would like to have you get here as soon as you can and that is with in three weeks.

Very truly yours,

John T. Benson

CREATOR OF THE STRANGEST FARM ON EARTH

BIGGER THAN A ZOO
BETTER THAN A CIRCUS

JTB:S

The place is a lot bigger than when you left

BIGGER THAN A ZOO
BETTER THAN A CIRCUS

This letter was sent by John T. Benson to Joseph Arcaris in 1940. Arcaris joined Benson, but when the park closed in 1942, he joined the Clyde Beatty Circus and later presented wild animal acts at Beatty's unique jungle tourist attraction in Fort Lauderdale, Florida. In the late 1940s, Arcaris returned to Benson's, where he entertained until his retirement in 1979. (Courtesy Joseph Arcaris.)

The lion wedding ceremony was a popular, one-of-a-kind wild animal act performed by trainer Joe Arcaris. The male lion was dressed in an outfit comprised of a jacket, top hat, and oversized eyeglasses, while the female wore a dress, necklace, and bonnet. The "groom" would then shake hands with the preacher as he smoked a pipe, and the "bride" would reward Arcaris with a kiss. A table and chairs were then placed inside the cage, where the lions sat as Arcaris fed the couple a wedding banquet. (Courtesy Joseph Arcaris.)

Trainer Arcaris thrilled the crowd as he guided a tiger through a hoop of fire. He was considered one of the best animal trainers in the country. He had worked with Clyde Beatty several times and took over the Beatty act to finish the season when Beatty's wife, Harriett, passed away.

One tiger jumps over the back of another tiger in the act trained by Joe Arcaris. Although Beatty featured a fighting wild animal act, Arcaris was quite different as he entered the arena with nothing more than a buggy stick.

According to trainer Joe Arcaris, his handstand on the backs of two lions was so unique and dangerous that other wild animal trainers would never attempt it. He had to approach and leave from the backs of the lions. If he ever lost his balance and fell over their heads, their jungle instincts would have them pouncing in response to what they would believe was an attack.

Another of the many Arcaris acts presented over the years was chimpanzees dressed in costumes. At Benson's, Arcaris trained various animals, including lions, tigers, bears, leopards, monkeys, elephants, pumas, chimpanzees, and even a 16-foot-long python.

Chimpanzees impersonating the Lone Ranger and Tonto always received a long and loud applause.

Six

ARTHUR PROVENCHER ERA

On February 28, 1979, it was announced that Arthur Provencher had purchased Benson's Wild Animal Farm. A Nashua native, Provencher owned a successful industrial park and truck leasing business in nearby Merrimack. At the industrial park, he had established a small free zoo that included Taco, his pet miniature mule, a selection of farm animals, a giraffe, and an elephant he acquired from a traveling circus.

For years, Provencher had noticed the gradual decline of Benson's, and after the final owner died, he started negotiations that took three years before being finalized. Consultants were hired to aid Provencher in redesigning and cleaning up the farm, while expanding the attractions into a complete family park. A full-time zoologist was also hired. Graphics providing information about the animals and their native habitats were installed outside the pens and corrals. Eventually Benson's reached the point of featuring more animals than both the Boston-area Stone and Franklin Park Zoos combined.

Many circus acts were booked, including the Wallenda high wire walkers, Philip Anthony's liberty horses, Rix Bears, and Frank Mogyorosi's Bengal tiger act. These were in addition to Larry Record's elephants, seals, horses, wild animals, and exotic birds. A lively parade wound its way around the grounds on a daily basis, often featuring local high school bands.

Each season Provencher poured money into the park with hopes of making it a profitable operation. He added exhibits, oversized storybook characters, flower bed plantings, a new gift shop, large arcade, theme restaurant, model railroad display, and a chain saw artist who created large attractive animals out of tree trunks.

Marketing, advertising, and promotion efforts were excellent with extensive newspaper, television, and radio advertising placed throughout the Boston area. There were many tie-in promotions with fast food chains, supermarkets, and name-brand products.

Attendance increased, but so did expenses. It was also a time when long-term financing was not available and only short term-loans were the alternative. For the 1987 season, the rights to use Terrytoon characters, including Mighty Mouse, were leased, and the name of the park was changed to New England Playworld. Unfortunately, the anticipated increase in attendance and revenue did not materialize, and the park closed for good at the end of the 1987 season.

This postcard shows the Lucky Elephant framed by the trellis with the windmill building to its right. For decades, visitors tossed coins onto the elephant's back. If the coins remained on its back, the person who tossed the coin was said to enjoy good luck in the future.

A mother and children were photographed after the elephant's trunk was changed to a raised one. The coins on the elephant's back were donated to various charities over the years. The first recipient was the Underprivileged Children's Fund, followed by the March of Dimes, and other worthwhile causes.

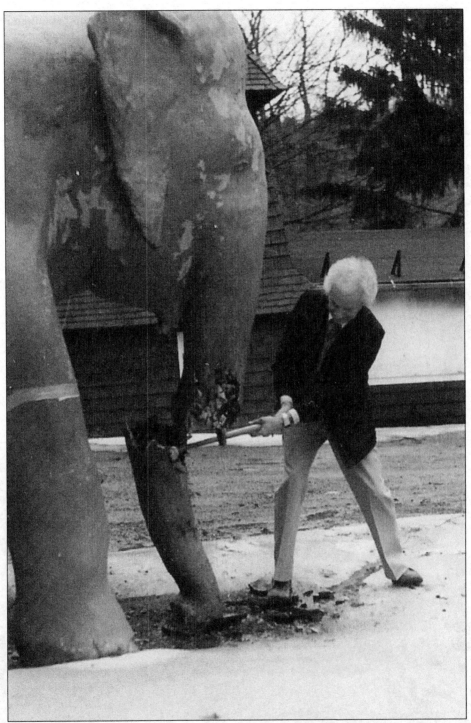

After purchasing the park, Arthur Provencher felt the Lucky Elephant could not bring good luck as its trunk was down. According to circus lore, only a raised trunk could bring good luck. Provencher took matters into his own hands and broke off the original trunk and then had it replaced with a raised trunk. (Courtesy Arthur Provencher.)

Provencher brought his pet miniature mule Taco, along with his giraffe, elephant, and other animals to their new home in Hudson. Taco died in 2008 at age 43.

Arthur Provencher greets one of his elephants during his first year as owner of Benson's Wild Animal Farm. Provencher had always been an animal lover who provided a free zoo at his industrial park in Merrimack, New Hampshire.

Provencher admires his newly installed C. P. Huntington train, which replaced the old, worn-out *Jungle Flyer* train. This was the first of many new rides that would be installed. When the park closed, the train was sold to the Great Escape amusement park in Lake George, New York, for use at Fantasy Island Amusement Park, its sister park on Long Beach Island, New York.

Here is a tree-shaded and nicely landscaped area of Benson's during the Provencher years. One of the major improvements Provencher began was to hire landscapers to redo the entire property with new plantings of shrubs, flowers, and trees.

Andrew Robel enjoyed watching the spotted deer in their pen. The Robel family attended the attraction as they drove from their home in Westchesterfield, New Hampshire, which is on the other side of the state near the Vermont border. (Courtesy George and Wendy Robel.)

Andrew Robel and his older brother Daniel were thrilled to pet the two special draft horses during a 1987 visit. (Courtesy George and Wendy Robel.)

Whenever possible, owner Provencher rode the lead elephant in the daily parade. Elephant trainer Bret Bronson leads Liz as costumed characters walk beside the trio. (Courtesy Arthur Provencher.)

One of the major attractions was Tony the gorilla. Provencher felt that the name Tony was definitely not a name for such a magnificent creature, and like a modern-day P. T. Barnum, he changed Tony's name to Colossus. (Courtesy Arthur Provencher.)

Colossus, a 500-pound silverback gorilla, admires one of his presidential campaign posters. During the time when political hopefuls were wooing New Hampshire primary voters, Provencher announced the candidacy of Colossus G. Benson for president. As the gorilla was too massive to transport, a chimpanzee claimed to be the campaign manager was taken to Concord for the filing of necessary papers. This Barnum-style action brought tremendous national publicity to Benson's. Following the park's closing, Colossus lived in a Florida zoo and in 1993 was traded to the Cincinnati Zoo. In 2006, at the age of 40, he passed away while undergoing dental surgery. (Courtesy Arthur Provencher.)

Two wooden soldiers stood guard at the elephant barn during the 1980 and 1981 Christmas seasons. The park was decorated with 100,000 lights and decorations. (Courtesy Arthur Provencher.)

To obtain many of the lights and decorations needed for the Christmas promotion, Provencher purchased the complete inventory of a holiday decorating company. The decorated boat on the lake was a beautiful sight. (Courtesy Arthur Provencher.)

Although over 35,000 visitors enjoyed the lights the first year and 50,000 the second, the break-even attendance numbers were never reached and the pageant was discontinued, primarily due to 24 percent interest rates in the early 1980s. (Courtesy Arthur Provencher.)

New England snow with Christmas lights and decorations presented a wonderful Christmas experience in 1980 and 1981. (Courtesy Arthur Provencher.)

The Fly-O-Plane ride with Red Baron markings on the body was an old ride that dated back several decades, but it nevertheless became a very popular ride at the park.

A colorful balloon ride had been manufactured only a year or two earlier, and Provencher paid a large sum of money to obtain this great favorite of teenagers.

The $250,000 Galleon ride was introduced to visitors in 1986. At that point, Provencher felt he had to install major rides to compete with nearby Canobie Lake and Whalom Park amusement parks.

The medium-sized 20-car family Fire Fly roller coaster was added in the early 1980s. It still operates at Santa's Village in Jefferson, New Hampshire.

Many new games were also introduced by Provencher in an effort to keep park patrons of every age group busy. This particular game was called the Frog Fishing Pond.

Seventeen units of the longtime arcade game favorite Skee-Ball were added to the many games available in the arcade. Other midway games included a shooting gallery, lollypop game, ping pong toss, basketball, and a dart game.

Pedal boats were constantly in demand during nice days when a leisure trip around Swan Lake could be enjoyed. Additional rides included the bumper cars, Tilt-A-Whirl, Looper, Tip Top, No. 16 Eli Ferris wheel, and a 1975 Jeep Renegade.

The turtle ride brought smiles to the faces of little tots who were too young for most of the other rides. Provencher purchased several additional kiddie rides to satisfy the little tots.

During the 61 years of Benson history, the elephant ride remained the most popular of the live animal rides. For many years it was Betsy who provided the ride, and in the final days it was Queenie and Liz.

A large model railroad and miniature circus exhibit was added in the mid-1980s. Always a circus fan, Provencher purchased a model circus and combined it with a model railroad, which had great appeal for all family members.

Cindy Martin Provencher performed with the elephants, whom she still refers to as her girls. Cindy is still very active with the group working for the completion of the Benson property as a town park. (Courtesy Cindy Martin Provencher.)

Cindy Martin Provencher, an all-around performer and animal attendant, enjoyed playing with the lion cubs in 1979. During her years at the park, she worked at just about every animal job. (Courtesy Cindy Martin Provencher.)

Following their morning swim in Swan Lake, Queenie and Liz would pause for the early camera-bug visitors. In this photograph they are doing what is known in the circus as the long mount. (Courtesy Brian Lawrence.)

An aerial photograph shows a portion of the 5,200 parking spaces in the huge parking lot. Several Hudson residents recall playing a game on Sunday afternoons when they would walk through the parking lot and list the names of state license plates. Some claim there were days when 48 different states were counted. (Courtesy Arthur Provencher.)

Swan Lake, with its charming fountain, always seemed to be a serene area despite the noise generated by large crowds. The fountain equipment is still submerged in the lake, and a future project is to restore it to working order.

This was a sign produced in-house that was displayed near the park's front entrance in 1987. (Courtesy Bret Bronson.)

With the Budweiser brewery located in nearby Merrimack, New Hampshire, it was not unusual to see a small version of the big beer wagons on the grounds. Fresh flowers were sold from this wagon.

In the spring of 1981, this group carefully moved the alligators from their winter holdings to the summer exhibit area. From left to right are Bruce Ashmore, Randy Warren, Tim Dutton, and Mika Nurmicko. (Courtesy Bret Bronson.)

The John T. Benson home was well kept during the Provencher era, as shown in this 1987 photograph. A portion of the well-landscaped property with beautiful and colorful flower beds is also visible. For an unknown reason, the state demolished the historic home after taking title to the property. (Courtesy Arthur Provencher.)

Jimmy Guill (left) and Larry Records stand ready with the elephants before starting one of their daily performances. (Courtesy Arthur Provencher.)

Bret Bronson guides Jackie, the new African baby elephant in a 1981 daily parade. Bronson was an elephant trainer at the park during the Provencher years. Although he now resides in Virginia, he still is active with the Benson group. (Courtesy Bret Bronson.)

Six years later in 1987, Bronson stands atop the much larger Jackie. Jackie, an African elephant, arrived at the park as a baby and grew quite large over the following six years. When the park closed, Al Jones of Hanover bought Jackie and leased her to Bronson, who formed an African elephant act that toured for a number of years. (Courtesy Bret Bronson.)

This was the hallway inside the elephant barn with the newly built nocturnal animal exhibits. With the new exhibit, visitors were able to observe nocturnal animals that could not have been seen under normal conditions. (Courtesy Bret Bronson.)

The outdoor lion exhibit area was built and opened in 1981. The giraffe and its barn are visible in the background. (Courtesy Bret Bronson.)

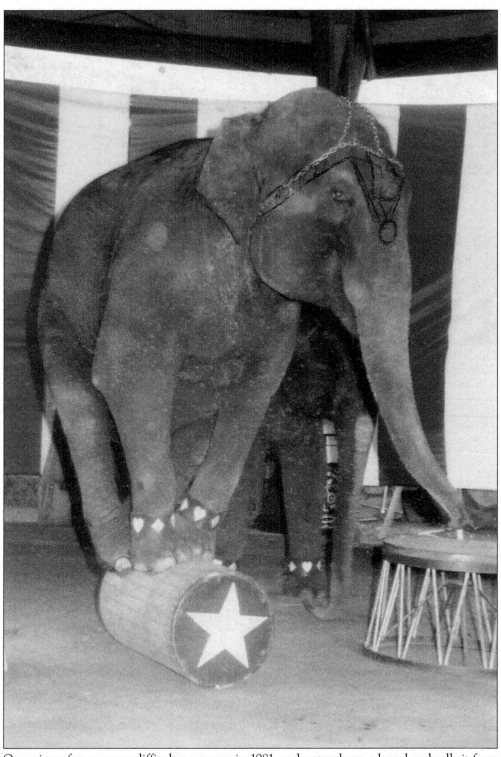

Queenie performs a very difficult maneuver in 1981 as she stands on a barrel and rolls it from tub to tub. (Courtesy Bret Bronson.)

The cover of the brochure for the 1987 season announced that the name of Benson's Wild Animal Park was changed to New England Playworld Amusement Park and Zoo.

Mighty Mouse, Deputy Daug, and other Terrytoon characters joined in-house character Benny Benson as they performed on the stage during one of several daily shows.

Mighty Mouse sits on a fiberglass elephant used during the 1987 season for parades and promotions.

Liz was enjoying some playtime when she climbed up on the embankment of a small pond on the property. (Courtesy Bret Bronson.)

This "Final Weekend" advertisement in the Nashua Telegraph broke the sad news to Hudson area residents that the park would close for good on October 12, 1987. The big two-day auction of animals, rides, games, and equipment was scheduled for Friday and Saturday, October 23 and 24.

Seven

BENSON'S LIVES ON

Following the Benson closing in 1987, its 166 acres gradually grew into a jungle-like setting, buildings fell into disrepair, graffiti was sprayed everywhere, and even the historic barn became a victim of arson. The New Hampshire Department of Transportation bought the property for $4 million, for wetland mitigation in connection with a circumferential highway, which never materialized. For 22 years, the land remained idle.

The idea of Hudson buying the land began in the 1990s when the town started putting money aside for its eventual purchase. In 2002, the state offered the property to the town for a bargain price of $188,000 on the condition that Hudson had to agree not to develop the site and ensure passive recreation use without disrupting the natural area. The sale was held up for a number of years, but on January 23, 2009, the transfer was made official during a statehouse ceremony in Concord hosted by Gov. John Lynch.

Esther McGraw served as chairman of the Benson's committee for nine years. The clearing and cleaning of the property began and continued under her direction. McGraw recalls that she even helped to build a small bridge. Over 200 hardworking volunteers cleared the overgrown brush, painted over graffiti, cleaned the grounds of trash, and other necessary tasks.

Hudson's highway department spread loam, planted seed, and continues to provide ongoing landscaping and grass-cutting, and Continental Paving repaved walking trails. Peter Ripaldi recruited Cub Scouts to clean the graffiti off the Lady Who Lived In A Shoe structure while his construction company repaired the broken side, reinforced the toe, installed Plexiglas windows, and refurbished the interior. After nine years of work, the Benson's committee was restructured by the board of selectmen with Harry Schibanoff appointed chairman. With board approval, the Town of Hudson then paid for new roofs on the existing buildings remaining on the property.

A new private, nonprofit corporation, Friends of Benson Park, formed under the leadership of Kenneth Matthews, has been authorized to raise funds for park improvements through memberships, donations, and fund-raising events. Benson Park is now open to the public with a variety of picturesque walking and bicycle paths, fishing in Swan Lake, picnic areas, and a Butterfly Garden. It is planned from donations, grants, and other sources to finance a variety of new projects such as a dog park, bandstand, senior citizen center, and a museum for Benson memorabilia.

With snow still on the ground, volunteer groups headed off into the undergrowth to start clearing the brush. Each group was assigned to a specific area. (Courtesy Esther McGraw.)

By 2010, an attractive new sign was erected at the present main entrance to the park. It is very visible and helpful in finding the new parking area, which is much smaller and located in a different area from the former parking lot entrance.

The building attached to the A-frame had deteriorated badly over the years and could not be saved. It became necessary to tear it down.

By 2010, the A-frame itself had received a new roof and paint job. It now serves as a picnic shelter.

The old depot train of the Worchester, Nashua, and Portland Division of the Boston and Maine Railroad stood for years on the other side of the road in need of considerable restoration. It was finally moved to the Benson property.

With some external repairs, the depot is ready to be moved to a permanent location where additional interior and exterior work will commence. At present, the use for this historic building has not been determined.

This is one of the many tranquil paths throughout the property that have been cleared by volunteer workers. The trails are well marked and vary in length.

Swan Lake has been restored to its original beauty. Future plans call for reactivating the fountain, which remains just below the water line.

This view was taken looking down Benson Boulevard in the 1980s, when the park was still operating. The gorilla house is on the right. (Courtesy Bret Bronson.)

This is the same location photographed in 2010. The gorilla house remains on the right, with a new roof. No decision has been made as to the use of this building in the future.

In the spring of 2010, the only remaining ticket booth at the park was in great need of repair.

By mid-summer 2010, the booth was moved to a permanent location where it will serve as an information kiosk. Its repair work is a project undertaken by Hudson Eagle Scouts.

The log building was used as an office by both John T. Benson and Arthur Provencher. During the consortium ownership, it was used as a gift shop. It has been mentioned that eventually it could become a caretaker's building. The lower brick building served as the kitchen for the beer garden and restaurant.

This new rest area was created for people who enjoy walking through the park. It is very restful and a nice place to sit in the shade. New benches are planned to be placed throughout the park.

During the years the property remained idle, vandalism had taken its toll on the Lady Who Lived In A Shoe unit. The area was cleaned, restored, and painted by Cub Scout Pack 20, assisted by Peter Ripaldi and his construction company. While in high school, Ripaldi was one-half of the unicycle-riding clowns Ham 'N Eggs. Additional Cub, Boy, and Eagle Scout projects are in the planning stages.

The old elephant barn, which also housed lions and tigers, is shown as it appears today. It is anticipated that after considerable interior work is completed it will house the Benson's Wild Animal Farm Museum, filled with memorabilia gathered over its 60-year existence.

The old elephant barn sits on a hill with a beautiful lawn leading up to it. The Hudson Highway Department keeps the grass well mowed and the grounds well landscaped.

This is one side of the Haselton barn. It has a new roof but needs extensive interior and exterior repair. The cupola has been removed for repair.

Large exterior patch areas are visible on the opposite side of the barn. The barn is not in the area used for the animal park. During the Provencher era, it was used by the maintenance department. The future use of the barn has not been decided at the time of this writing.

The large, fancy cupola awaits repair as it sits on the ground surrounded by tree and brush overgrowth. When repaired and painted, it will be placed back at its position on the top of the barn.

Eight

BENSON MEMORIES AND MEMORABILIA

Benson's memorabilia can be found in all sizes and shapes in many places, not just in Hudson. A good portion remains in the collection of former owner Arthur Provencher of Henniker, New Hampshire. The two Bostock and Wombwell circus wagons now reside at the Circus World Museum in Baraboo, Wisconsin,

Many private collections are still active throughout the area, and a check of e-Bay always seems to find items for sale. Photographs, postcards, and souvenirs, along with family photographs and home movies taken at the fun spot, are still cherished in Hudson-Nashua homes. Ruth Parker of the Hudson Historical Society has put together an interesting historical film show that is presented to the public from time to time.

The park's roller coaster now operates at Santa's Village in Jefferson, New Hampshire, and the Galleon ship ride can be found at Knoeblels Grove Amusement Park in Leysburg, Pennsylvania. It is believed that the four-room house carved out of a giant redwood tree resides in a Florida Ripley's museum. A large fiberglass bull, mounted on a trailer, was last seen at a car dealership in Salem, New Hampshire. Seven buildings, including the Lady Who Lived In A Shoe, have been saved and are in various stages of restoration on the grounds. Two concrete lions still repose on the lawn of the Hudson home of Maurice and Carol Viens.

In 1995, the Hudson Chamber of Commerce purchased five of the large wooden soldiers and presented them to the town of Hudson. They were repaired and painted by students and can now be seen on the town common each Christmas season. The Friends of Benson Park organization has intentions of eventually filling a museum at the Benson Park with memorabilia donated from many sources.

An editorial appeared in the June 20, 2010, *Telegraph* newspaper that read, "The people who volunteered their time to revamp the property have given a different kind of gift to people with fond memories of Benson's Wild Animal Farm, a piece of their childhood back."

This is a scene in 2009, when former owner Arthur Provencher presented his extensive memorabilia in the old Hudson library. During two weeks in July and four days during Hudson's Old Homes Days, over 4,000 people visited the exhibit.

A miniature mock-up of Benson's produced during Provencher's ownership was on display under an acrylic dome with a complete legend of its contents.

This partial view of the Provencher exhibit was taken from the center of the large room. Artifacts and memorabilia are visible throughout the area.

This is another view of the extensive exhibit taken in the opposite direction. Arthur Provencher also showed a number of videos of the park.

The Bostock and Wombwell's menagerie living wagon, built in 1883, served as an office and living quarters for owner Edward H. Bostock in England.

John T. Benson brought the wagon to Hudson in 1932 after Bostock went out of business. A roof was built over the band and living wagons to protect the wagons from the elements. Stairs and a deck were built in front of the living wagon so visitors could observe the interior furnishings and office.

The living wagon was purchased by a private bidder at the auction and was later donated to the Circus World Museum in Baraboo, Wisconsin. The two photographs on this page show the wagon when it was being rebuilt and renovated at the museum.

The Bostock and Wombwell bandwagon, the oldest known English bandwagon in existence, was on display for decades at Benson's. It was purchased at the auction by the Circus World Museum in Baraboo, Wisconsin.

Each year, the Circus World Museum loaded the wagon on its circus train, which transported it to Milwaukee, Wisconsin, to appear in the Great Circus Parade.

The wagon, pulled by a team of horses and carrying a circus band, made its way along the parade route in the 1990s. From all indications it appears that the Great Circus Parade will not be held during the next few years, due to money problems.

The Circus World Museum rents its wagons out for parades, fairs, movies, and various other projects. This photograph was taken in a Wisconsin town when the former Benson bandwagon was used to carry Santa Claus in a Christmas parade.

Many of the circus wagons were lined up early on the morning of the Milwaukee parade in the 1990s, as they awaited the horse teams assigned to pull them along the streets of Milwaukee.

As he visited Milwaukee for the circus parade in the 1990s, former Benson's owner Arthur Provencher was reunited with his bandwagon; his park was only a memory.

New England Playworld in 1987 was backed with heavy advertising. These two promotion pieces featured Mighty Mouse and the Terrytoon characters and were distributed throughout the area.

A worn and tattered wooden soldier and a prehistoric monster in need of repair stood in a New Hampshire field a few years ago. Both are now under cover.

In 1987, circus model builder Leo Metzger constructed this model bandwagon, which still remains in the Arthur Provencher collection. It was part of the memorabilia on display during the 2009 exhibit at the old Hudson library.

Nine glass display windows were built into a wall of the elephant barn in the 1980s, allowing visitors a close look at the various reptiles on display. The entire wall can be found today at York's Wild Animal Kingdom in York Beach, Maine. (Courtesy Bret Bronson.)

A variety of patches were worn on uniforms and shirts during the many years the attraction operated. This patch is from the Arthur Provencher era.

BENSO

WILD ANIMAL PARK
Rte. 111 Hudson, N. H.

This colorful map of the park was printed in 1980 and remains in the collection of many Benson fans. During the 2010 Hudson Old Home Days, a framed map sold at silent auction for $140. A similar black-and-white version of the map hangs on the wall of a Nashua restaurant.

Robert Hall of Bangor, Maine, sold this circus wagon to Arthur Provencher in 1979. It remains in the former owner's possession and was on display outside the Hudson library during his 2009 exhibit.

Car stickers have always been popular, and Benson's gift shop sold many styles over the years. Most of the stickers or decals featured wild animals. (Courtesy Esther McGraw.)

Mary was a cassowary, a large flightless bird from New Guinea. She lived to be almost 60 years old at Benson's, becoming quite temperamental in her later years. When she passed away, employees decided to have her stuffed to allow her legend to live on. Her neck is a brilliant blue with vertical streaks of bright red below. She was on display at the 2009 Benson's exhibit at the old Hudson library.

The four-room log home was a good attraction during its time at Benson's and is now on display at the Ripley's Museum in St. Augustine, Florida.

The trailer-mounted large fiberglass bull is often seen at automobile dealers in the area.

The roller coaster still thrills riders at Santa's Village in northern New Hampshire.

Roland Tiebor's seals and sea lions played tunes on these musical horns, which remain in their traveling case in the possession of Esther McGraw. The song played at the end of their act was "My Country Tis of Thee." (Courtesy Esther McGraw.)

This 13-inch special wood type plate dates back to the John T. Benson days. (Courtesy Esther McGrew.)

This is one of the many pennants sold over the years at the Benson gift shops. There were many styles and colors sold that now appear in private collections. (Courtesy Hudson Historical Society.)

Former concrete lions from Benson's are now part of a very attractive display in the front yard of Maurice and Carol Viens of Hudson. (Courtesy Maurice and Carol Viens.)

Once the stub was torn off the end of the Provencher-era tickets, they became a colorful, treasured bookmark. Child tickets were blue, and adult tickets were red.

Every Christmas season, these five wooden soldiers stand at attention on the town common. They are maintained by the town of Hudson.

This is an old bumper sticker from a private Benson's collection. Although this is listed as a sticker, they were heavier advertisements that were attached to automobile bumpers with two pieces of wire. (Courtesy Bret Bronson.)

Former Benson's office staff and zookeepers gathered with Arthur Provencher on June 27, 2009. The group walked the grounds and then posed for this photograph in front of the gorilla building. It was the first time that Provencher had set foot in the site since he left the park after its closing. Pictured from left to right are Bret Bronson, Chick Forrence, Rhonda (Perry) Knowlton, Bruce Ashmore, Kathy (Weldon) Mills, Cathy Provencher, Jodi Provencher, Rachel (Comtois) Feeley, Gary Provencher, Cindy Provencher, Cathy Knox, Tim Dutton, Jay Knox, Steve Klein, Kevin Houle, Catherine Kenney, and Arthur Provencher. (Courtesy Cindy Provencher.)

CPSIA information can be obtained
at www.ICGtesting.com
Printed in the USA
LVOW04*2006290118

564439LV00030B/605/P